U0199830

故園畫憶

庚寅中秋

韩启德题

故园画忆系列

Memory of the Old Home in Sketches

西宁掠影

Sketches of Xining

苏繁昌　绘画 撰文

Sketches & Notes by Su Fanchang

學苑出版社

Academy Press

图书在版编目(CIP)数据

西宁掠影 / 苏繁昌绘画撰文. — 北京 : 学苑出版社，2019.4
(故园画忆系列)
ISBN 978-7-5077-5672-2

Ⅰ. ①西… Ⅱ. ①苏… Ⅲ. ①建筑画 - 作品集 - 中国 - 现代②西宁 - 概况 Ⅳ. ①TU204.132②K924.41

中国版本图书馆CIP数据核字(2019)第051733号

出 版 人：孟 白
责任编辑：周 鼎 康 妮
出版发行：学苑出版社
社　　址：北京市丰台区南方庄2号院1号楼
邮政编码：100079
网　　址：www.book001.com
电子信箱：xueyuanpress@163.com
联系电话：010-67601101(营销部)、67603091(总编室)
经　　销：全国新华书店
印 刷 厂：河北赛文印刷有限公司
开本尺寸：889 × 1194 1/24
印　　张：5.5
字　　数：150千字
图　　幅：103幅
版　　次：2019年4月北京第1版
印　　次：2019年4月北京第1次印刷
定　　价：45.00元

目　录

民风民俗

青海自然风光

Contents

Customs and Cultures

Natural Scenery

序

一座城市的山川景物往往体现着它卓尔不群的魅力，青海省的省会西宁就是这样一个地域广博、文化缤纷、多民族合睦相处的高原城市。

近现代以来，青海的妖娆与神韵，吸引着国内外众多艺术家寻幽探奇，也留下了他们对这一方土地的绘画印迹。青年画家苏繁昌就是这些艺术探求者中孜孜以求的一员。

为了展示西宁这座城市的文化价值和魅力所在，在相当一段时间内，繁昌以他极高的艺术热忱，为挖掘西宁乃至整个青海文化全貌，春来暑往，走遍古城的大街小巷，进行艺术采风和绘画写生。他以一个西部艺术家特有的敏锐和严谨的画风，历经数年完成了这部作品。我与繁昌虽然相交多年，但当他一张张速写递到我眼前时，让我感到欣慰的是这些作品不仅仅是通过画家所描绘的对象来展现高原夏都的场景，更多的是透过这些作品，体现了繁昌日渐成熟的艺术风貌和作品强烈的文化感染力，它们所散发出来的人文气息和艺术价值就目前现有此类速写作品而言，我认为是不可估量的。其一，繁昌的速写更注重“写”的意识，用中国画的笔墨意味来表现所描绘的对象，这往往是他的作品与那些带着强烈明暗、素描味道极强的建筑图所不同的地方。其二，与他那些大气磅礴的山水画相比，这些纸不盈尺的速写，在熟知他的人看来似乎是小菜一碟。但是，这些作品内含的小画面透露着大气场，小速写呈现着大场景。其表现内容极为丰富，是集建筑、宗教、民俗、历史、人文于一刊的鸿篇巨制。这或许恰恰是他的过人之处，也正是他独具匠心的地方。此前以往，看到过诸多这类出版的书籍，其内容大多拘于以建筑尤其现代建筑来展现一个城市的面貌。而这本书，繁昌以娴熟的技艺和独特的构思，再加上新颖的题材，使之成为这一类出版物的新亮点，他已经从建筑速写的规范内脱颖而出。因为，在繁昌的速写之中既有描绘土族波波会，跳神、祈福；藏传佛教的晒佛、藏戏；旧城遗迹、古墓道观、街景小吃等等。该作品极为全面地诠释了青海西宁的璀璨文化，让观者耳目一新，感受到了西宁的独特魅力。或许还可以说：就繁昌的速写描绘内容而言，不仅仅是几座古旧的寺观、几根潇洒飘逸的线条，也不仅是几组姿态各异的人物、几张丰富多彩的场景，它们更多的是借笔墨抒发画家心灵所建构的宏大文化帷幕。从塔尔寺到湟水河，从青唐城到北山寺，从青海湖畔到贵德古城，画家的笔墨随历史之长度，从远到近，线条则从古到今，紧扣观者的眼球，进而一一展开……。我想，繁昌画作之中所谓的“大巧若拙”即是此情、此景、此画所仰承的更多的是一种文化，一种最民族、最深入人心的大美青海之文化意象。

繁昌自幼好画，先后求学于西安与京城。他性格憨厚朴实，其画作多有获奖。该速写集是繁昌为西宁乃至青海地域文化的艺术观照，观其作、品其画，可视为现今众多速写艺术作品的佳作，值得玩味，值得品鉴。

我国绘画名家成于速写者不在少数，速写无疑是他们名著世间的法宝之一。我想借此寄望繁昌能够循此路径，持之以恒，不断探求！

是为序。

王万成

西北民族大学美术学院副院长

甘肃省美术家协会副主席

Preface

The mountains, hills and landscapes of a city usually represent its beauty and uniqueness. Xining is a multi-ethnic and multi-cultural capital city on the Qinghai-Tibet Plateau.

Since contemporary and modern times, Qinghai has attracted many artists from abroad and China who have created many artworks about this land. Su Fanchang, a young painter, is one of these art explorers.

In order to display the culture and beauty of Xining, for a long time, Fanchang has been visiting the streets and alleys of this ancient city dedicating himself to sketching with his pens and his passion for art. He finished these artworks with his artistic acuteness and preciseness, rendering both visual enjoyment and cultural inspiration for our readers.

Fanchang has had a passion for drawing ever since he was young. He studied first in Chang'an and then in Beijing. He is a kind-hearted and honest person. Many of his artworks have won awards. This collection of his sketches is mainly from his recent works; we are sure they will be appreciated and appraised.

Many artists in China became famous because of their sketches; so, this can be a good way leading to success in the arts. Here I want to wish for him to follow his heart's desire and have a successful career through his dedication and persistence!

This can serve as the preface of this book.

Wang Wancheng
Vice President, College of Fine Arts,
Northwestern Ethnic University
Vice President, Gansu Provincial Artists' Association

古代遗存
Imperial Tombs Relics

青唐城

青唐城就在今天的西宁市区，唐代这里叫鄯城。安史之乱后，吐蕃从唐军手中夺取了鄯城。那时候，城池四周山上林木参天，青翠葱茏，所以吐蕃族称之为青唐城，如今只剩下残垣遗迹。

Qingtang City

Xining was also called Shan City during the Tang Dynasty(618-907). After the An-Shi Revolt, the Tibetan army seized Shan City from the Tang troops. Because the city was surrounded by a green forest, the Tibetans called it Qing Tang City (Green Tang City). Now only ruins remains.

沈那遗址

位于西宁市城北区小桥大街毛胜寺西台地上。沈那遗址以齐家文化居住遗存为主，还有少量的马家窑文化马家窑类型、半山类型和卡约文化遗存，是约 3500 年前的古羌人聚落村，是远古人类从新石器时代向青铜时代过渡的一种文化遗存。该遗址是我国迄今发现的面积较大、文化层堆积较厚、文化内涵相当丰富、保存现状较好的多种文化并存地点之一。

Shenna Site

Located on the Western Terrace of Maosheng Temple, Xiaoqiao Street, Chengbei District, the site is a settlement village of ancient Qiang people about 3500 years ago. It is a cultural relic of the transition from Neolithic to Bronze Ages. Shenna Site is one of the coexistence sites of many cultures with large area, thick accumulation of cultural layers, rich cultural connotations and good preservation status.

南凉虎台遗址·大门

位于西宁市西郊，为4世纪南凉遗址。原台共9层，台下可陈兵10万，台上用于军事检阅，现仅存土丘。登上虎台，可览西宁风光。

Nanliang Hutai Relics Park · Gate

Located in the western suburb of Xining, it is a 4th century relic from the Nan Liang Dynasty(397-414). Originally a nine–storey stronghold, the first floor could house about 10,000 soldiers. The top was a military outlook post. Now only soldier lookouts remaied.

南凉虎台遗址·土墩

土墩用于军事检阅，如今已经荒颓，只剩下土台一座，被包围在高大的现代化的楼群建筑之中。

Nanliang Hutai Relics Park · Hill

The mounts were previously used for military lookouts. Now only one desolate mount is left, surrounded by modern building complexes.

南凉康王墓

魏晋时期，世居塞北的鲜卑部落陆续南迁，西宁也在历史上第一次成了王国的都城，秃发三兄弟建立起的南凉王国给西宁留下了除却历代城墙以外最宏伟的历史遗迹——康王墓。

Tomb of Prince Kang Prince of the Nan Liang Dynasty

About 1800 years ago, during the Wei and Jin Dynasties (220-589), Xining became the capital of an empire for the first time. The Nan Liang Dynasty (397-414) established by the "Three Bold Brothers" left a grandiose historic relic in Xining—the Tomb of Prince Kang.

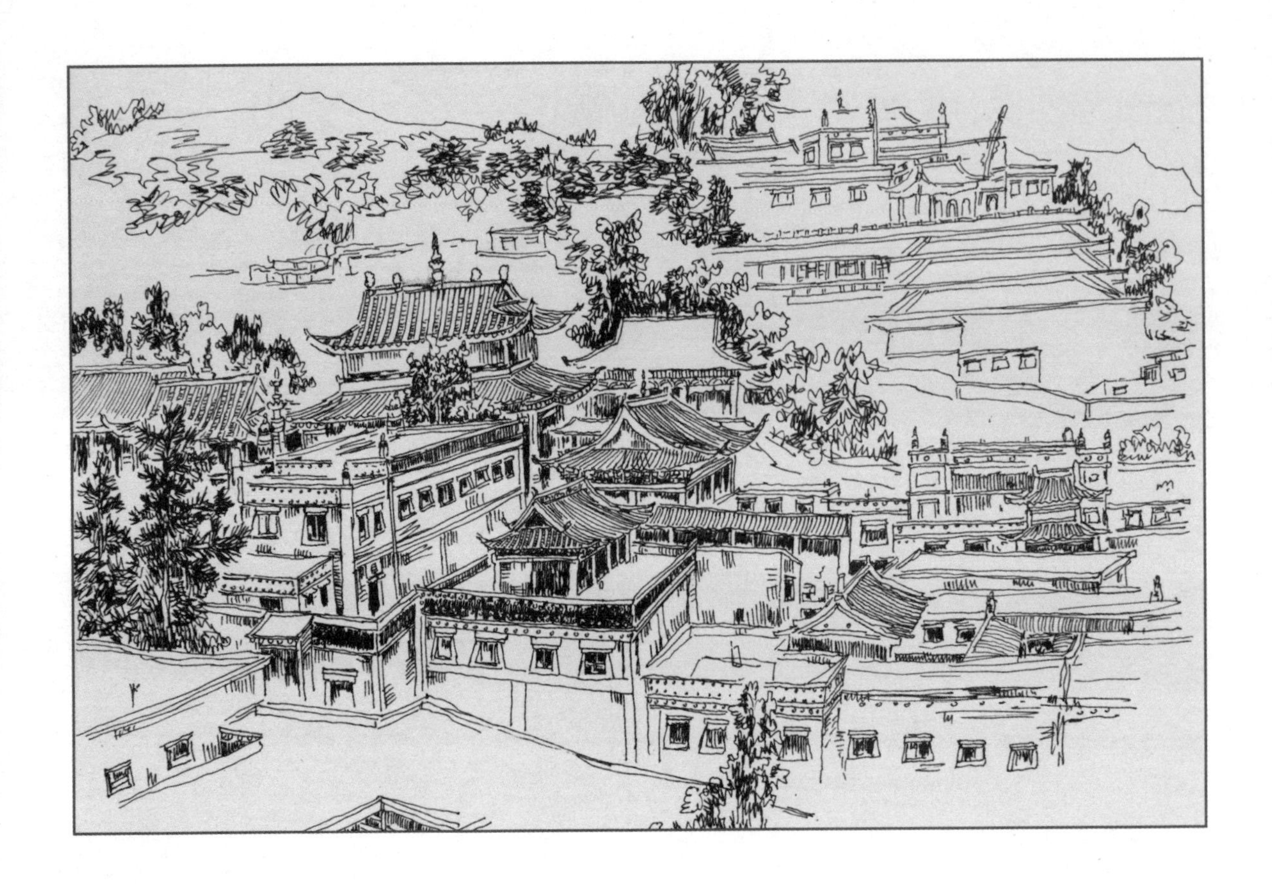

塔尔寺·全景

塔尔寺位于西宁市西南25千米处的湟中县县城鲁沙尔镇。塔尔寺共有大金瓦寺、小金瓦寺、花寺等大小共1000多座建筑，4500多间殿宇、僧舍，规模宏大。宫殿、佛堂、习经堂、寝宫、喇嘛居住的扎厦以及庭院交相辉映，浑然一体，自古以来即为黄教中心及佛教圣地。

Kumbum Monastery · Overview

There are more than 1,000 courtyards and 4,500 residences for monks in Kumbum Monastery, among which Great Golden Roof Hall, Small Golden Roof Hall and Hua Monastery are the most famous. It has been a holy place for the Yellow Hat Sect of Tibetan Buddhists for a long time.

塔尔寺·如来八塔

塔尔寺得名于大金瓦寺内为纪念黄教创始人宗喀巴而建的大银塔，为中国藏传佛教格鲁派（黄教）六大寺院之一。图为如来八塔。

Kumbum Monastery · Buddha Tower

Located at Lugar Village in Huangzhong County, 25km southwest of Xining, Qinghai. It derived its name from the Dayin Stupas in Great Golden Roof Hall. Kumbum Monstery is one of the six major monasteries of Gelugpa (Yellow Hat) Sect of Tibetan Buddhism.

塔尔寺·时轮经院

塔尔寺内的时轮经院是研习天文历算的学府。学习成绩优秀者，可被授予“泽然巴”格西学位。获得学位的僧人深受僧俗尊敬，在寺院享有很高的地位。

Kumbum Monastery · Monastic College

The Monastic College is for the study astronomy and arithmetic. Excellent students can be granted Geshe Shyrampa, the highest award for monks.

塔尔寺·大经堂

大经堂是塔尔寺中规模最大的土木结构的藏式平顶建筑。初建于明万历三十四年（1606 年），是寺院喇嘛集中诵经的地方，堂内设有佛团垫，可供千余喇嘛集体打坐诵经。内部陈设十分考究，有各式天花藻井饰有黄红绿蓝白五色的幡、帏，还有珍贵的大型堆绣，彩画细腻生动。

Kumbum Monastery · Great Hall

A large wooden Tibetan style structure built in 1606 during the Ming Dynasty, it is where monks study Buddhist scriptures. The Great Hall can hold a thousand monks sitting and praying together. Its interior layout is very delicate.

塔尔寺·藏经阁

塔尔寺的藏经阁于 2002 年开工建设，2007 年完工。建筑全部为木结构，分上下五层，第四层和第五层仿西藏大昭寺而建。

Kumbum Monastery · Depositary of Buddhist Texts

Its construction began in 2002 and was completed in 2007. The Repository is a five-storey wooden building. The fourth and fifth storeys are architectural copies of the Jokhang Monastery in Tibet.

塔尔寺 · 大金瓦寺

大金瓦殿位于塔尔寺正中。藏语称为"赛尔顿庆莫"，即金瓦的意思，是塔尔寺最为庄严、富丽的建筑之一。

Kumbum Monastery · Great Golden Roof Hall

Located at its center, it is the monastery's most solemn and splendid building. In Tibetan language it means "Golden Roof".

塔尔寺·酥油花

酥油花是一种用酥油塑形的艺术品，制作技艺特殊，为“塔尔寺三绝”（酥油花、壁画、堆绣）之一。酥油花艺术继承了藏传佛教艺术的精、繁、巧的特点，其设计、制作自古是师徒口手相传，一般都在封闭的环境里制作。

Kumbum Monastery · Butter Sculptures

As one of the Three Unique Art Forms in Kumbum Monastery, butter sculpture is a special art craft made of yak butter. This skillful handcraft of Tibetan Buddhism combines the art characteristics of delicacy, complexity and dexterity. Its design and production techniques are passed on by craftsmen through apprenticeships. Usually they are created is a very closed environment.

法幢寺

现位于西宁市南山公园南禅寺旁边，法幢寺是青海省最大的汉传佛教比丘尼寺。始建于1943年，2003年该寺从园树庄搬迁到现址。

Fazhuang Temple

Fazhuang Temple was built in 1943 next to Nanshan Temple in Nanshan Park, Xining. It is the largest Buddhist temple for nuns in Qinghai Province.

东关清真大寺·全景

位于西宁东关大街路南一侧，是西宁古城著名的建筑，也是我国西北地区大清真寺之一。

Dongguan Grand Mosque · Panorama

Located on Xining's Dongguan Street, this famous ancient building is one of the grand mosques in northwest China.

东关清真大寺·礼拜堂

东关清真大寺的礼拜堂占地面积1102平方米，雕梁彩檐，金碧辉煌，大殿内宽敞、高大、明亮，可以同时容纳3000多名穆斯林进行礼拜。殿内古朴雅致，庄严肃穆，富有浓郁的伊斯兰特色。

Dongguan Grand Mosque · Worship Hall

The 1102 km^2 grandiose, bright. hall accommodates 3,000 worshippers. The palace-like design, has typical Islamic characteristics.

东关清真大寺·邦克楼

邦克楼是清真寺群体建筑的组成部分之一。“邦克”为阿拉伯语音译，意为尖塔、高塔、望塔，即宣礼塔。又称为唤礼塔。中国穆斯林称其为邦克楼、望月楼。专门用作宣礼或确定斋戒月起讫日期观察新月，是清真寺建筑的标志之一。

Dongguan Grand Mosque · Bangke Hall

Part of the Islamic Mosque complex, bang ke is Chinese for the Arabic, meaning “calling to prayer”. It is one of the major symbolic buildings in a mosque.

凤凰台拱北·正门

位于西宁市南山公园内。1987 年重修，建成五级亭楼式拱北建筑，庄严肃穆，是河湟一带穆斯林最早的宗教活动场所，常有穆斯林前去拜谒。

Phoenix Qubbah · Front Door

Rebuilt in 1987 in Xining's Nanshan Park near Hehuang, the Qubbah is a five-storey Qubbah style worshipping place for Muslims.

凤凰台拱北 · 先贤墓

先贤墓坐北朝南，飞檐秀出，造工精细，为伊斯兰教的先贤陵墓。

Phoenix Qubbah · Ancestry Cemetery

This delicate Qubbah, situated in the north and facing south, is a cemetery for Islamic ancestors.

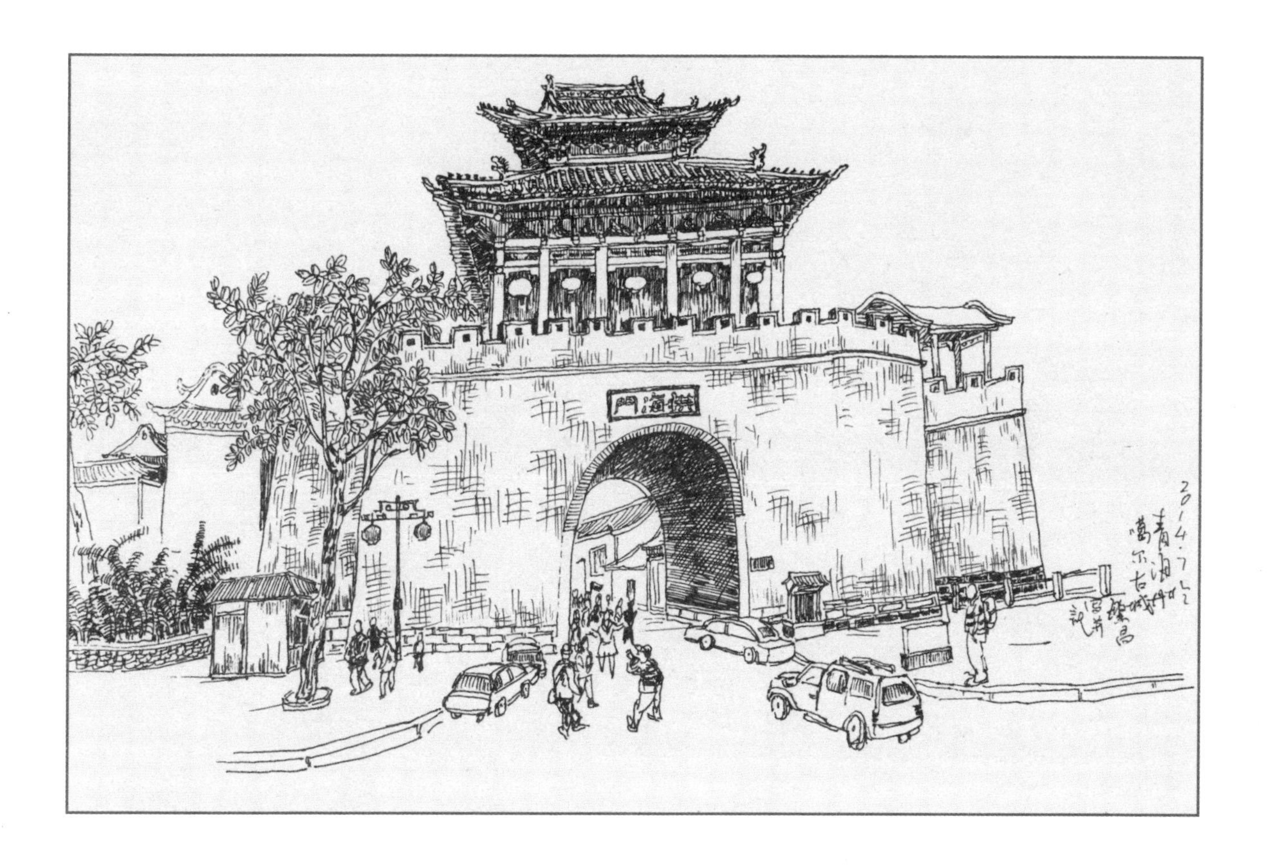

丹噶尔古城·拱海门

拱海门是丹噶尔古城的西城门。旧时去青海湖畔，丹噶尔古城是必经之地。古代拜祭青海湖的钦差大臣从东城门进、西城门出，在西城门外要举行简单的祭海仪式，因而称西城门为“拱海门”。

Ancient Dange'r · Gonghai Gate

It is the west city gate of ancient Dange'r and the only way towards Qinghai Lake. In ancient times, the imperial commissioners who came to worship Qinghai Lake entered from the East Gate and exited through the West Gate, where a simple worshipping ceremony was held honoring the Sea (Qinghai Lake), thus the West Gate is also called Gonghai Gate ("worshipping the Sea").

丹噶尔古城·城隍庙

位于丹噶尔古城内，在拱海门（西门）附近，始建于清乾隆四十一年（1776年）。有山门三间，有钟鼓楼阁、门楼戏台、正殿、牌房、花园、书房等。庙宇建筑宏伟，壁画精美，是西北地区保护最完整的城隍庙之一。

Ancient Dange'r · City God's Temple

Close to West Gate of Dange'r, it was built during the reign of Emperor Qianlong of the Qing Dynasty (1644-1911). This grand temple with beautiful frescos, is the best preserved City God Temple in northwest China.

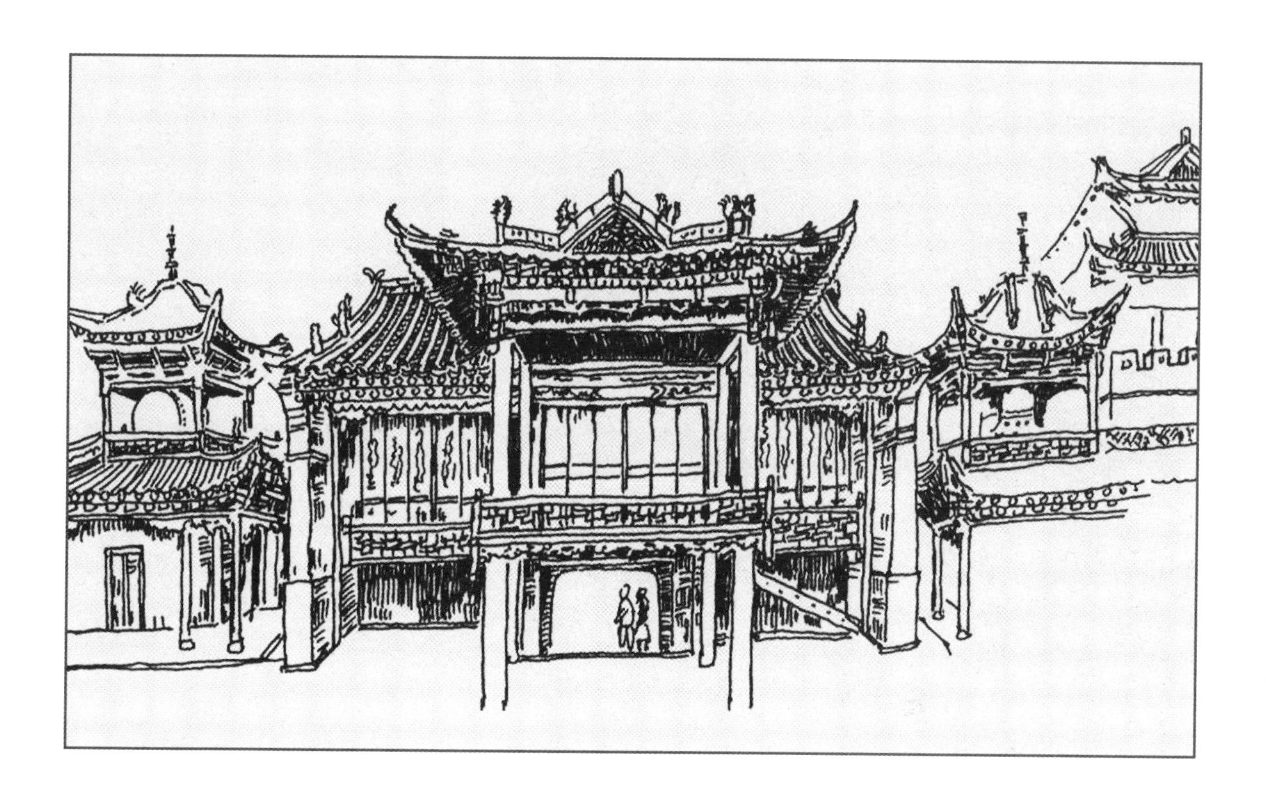

丹噶尔古城·城隍庙戏台

丹噶尔古城的城隍庙内有大戏台一座。戏台气势宏伟，刻镂精致，为民间艺术一绝。旧时农历正月十三至十六，都要组织民间艺人在此进行地方曲艺演出等活动。

Ancient Dange'r · City God's Temple Drama Stage

Founded in 1776, it is one of the most well preserved City God's Temple in the northwest region. Its stage is magnificent and engraved with delicacy. From the thirteenth to sixteenth day of the first lunar month, folk artists are organized to perform local folk arts, carry out city god worship and City God patrol activities, and display unique regional folk culture.

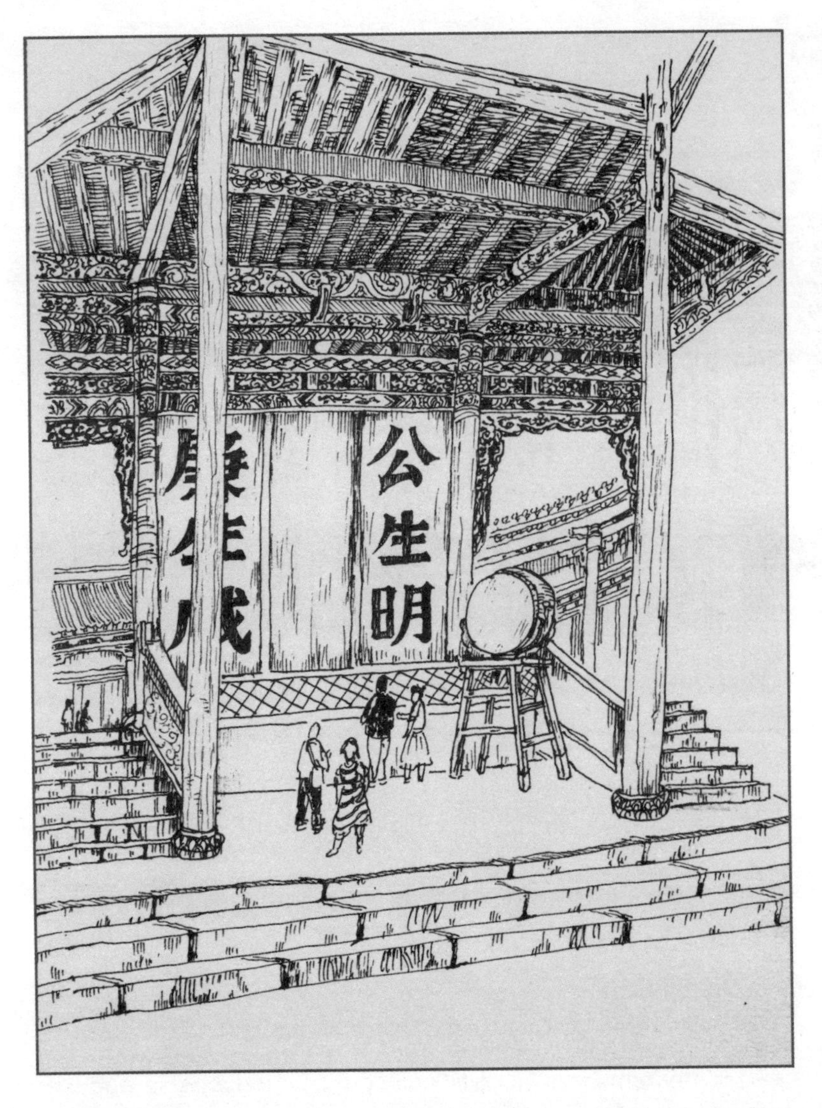

丹噶尔古城·厅署

清道光年间，因海藏通商，特于湟源设立丹噶尔厅，属西宁府。现修复的丹噶尔厅署为前后两院，前院有大堂及六房，后院有议事堂等。

Ancient Dange'r · City Hall

Due to its commercial importance of the city, during the reign of Daoguang Emperor of the Qing Dynasty (1644-1911), this city hall was established as an affiliated office of the Xining government.

丹噶尔古城·仁记商行

丹噶尔古城内现存的一座洋行，由英国人创办。它是旧中国外商在湟源经商的历史见证。清代嘉庆、道光年间，丹噶尔古城商贸鼎盛，欧美等国商人纷纷在此驻店经商，主要从事羊毛的收购与贩运。

Ancient Dange'r · Renji Commercial Firm

The existing foreign firm in Dange'r was established by the British. It is an historic witness to how foreign traders did business in Huangyuan before the Liberation.

土楼观·全景

又称“北禅寺”，位于西宁市北湟水之滨海拔 2400 多米的北山上。因北山山崖层叠，远眺似土台楼阁高高矗立，故又名土楼山，北禅寺也被叫作土楼观。北禅寺早先为佛教寺庙，是青海境内最早的宗教建筑，初建于北魏时期（386—534 年），距今已有一千多年。

Tulouguan · Panorama

Also called Beichan Temple, it is located at North Mountain (2400m above sea level) near the bank of North Huangshui River of Xining. Due to the cliffs of North Mountain, when seen from afar, it looks like a mount standing straight, thus the name Tulou Mountain. Beichan Temple is also called Tulouguan.

土楼观·大门

土楼山是一个很神秘的地方，集儒释道于一山，山上供着诸多的神仙，甚至连阎罗王也有供奉，由此更加增添了土楼观的神秘感。清代诗人张思宪曾做《北山烟雨》诗，“北山隐约书模糊，烟雨朝朝入画图。却忆草堂留我住，爱他水墨米颠呼”。图为土楼观的大门。

Tulouguan · Main Gate

Tuloumountain is a very mysterious place, which integrates Confucianism, Buddhism and Taoism on a mountain. There are many immortals on the mountain. Even King Yan Luo has been offered here, which adds to the mystery of Beishan Temple. The picture shows the main gate of Tulouguan.

土楼观 · 景观

土楼观是依土楼山特殊丹霞地貌造型而建造的。这里的岩石是紫红色的砂岩、砾岩，其间还夹有石膏和芒硝层。土楼观被称为“丝绸之路”南道的一颗明珠。

Tulouguan · View

Tulouguan was built based on the Danxie landform in Tulou Mountain. The rocks here are a mostly purplish red sandstone, conglomerate rock, mingled with parget and mirabilite. The architecture style and culture here are regarded as "Shining Pearls" on the southern Silk Road.

土楼观·悬空栈道

土楼山发育完好的丹霞地貌向里凹进，形成大小不等的洞穴。栈道回廊紧靠悬崖，甚至悬空架设，将殿宇楼阁与洞穴群相连，颇为壮观。

Tulouguan · Suspending Foot Planking

The well-developed Danxia landform cuts into the mountain, forming caves. A long, plank footpath of planks connects the pavilions and pagodas with the clusters of caves. The paths built against the cliffs and suspending in the air, create a spectacular view.

土楼观·悬空寺

始建于北魏时期（386—534年），是中国第二大悬空寺，多年来以其特有的景致吸引着人们的目光。大量的壁画，丰富的历史，从悬崖间展现在人们眼前。

Tulouguan · Suspended Temple

Built during the North Wei Dynasty (386-534), Suspended Temple has long attracted great attention because of its unique landscape. After thousands of years, it has developed a unique culture.

赞普林卡·正门

赞普林卡位于西宁的西大门湟源，距西宁仅43千米，赞普林卡总占地43亩，分前后两院，前院为佛殿，后院为藏式宾馆和园林。

Zamprinka · Front Gate

Zamprinka, located in Huangyuan, the west gate of Xining, is the intersection of the ancient Silk South Road, the Three Gorges, and the throat of the sea and Tibet. It is only 43 kilometers away from Xining, the capital of the province. The total area of Zamprinka is 43 mu, divided into two courtyards, the front courtyard is the Buddhist Hall, the back courtyard is the Tibetan Hotel and the Royal garden.

赞普林卡·主殿

为五层楼藏式风格建筑，内供奉有藏王松赞干布和王妃文成公主、尺尊公主的雕像。是汉藏民族团结的历史见证。

Zamprinka · Main Hall

This five-storey Tibetan style structure, with images of the Buddha, Tibetan King Songtsen Gampo, Princess Wencheng and Princess Chizun inside, marks the unification of Han and Zang ethnic groups and is known as the Holy Place of Xiadu.

拱辰门

位于西宁市城中区北大街最北端。拱辰门始建于明洪武十九年（1386 年）。西宁城呈方形，城东、南、西、北四面各开一门，四门均建有门楼。北门坡下又有北门泉，故北门俗称“水门”，城楼匾题“拱辰”。现建筑为近代在原址上重建。

Gongchen Gate

Located at northernmost North Street, Chengzhong District, Xining, it was built in 1386, during the reign of Hongwu Emperor in Ming Dynasty(1368-1644). Four gates East, South, West and North serve the city. At the foot of North Gate, there was a fountain, so North Gate was also called “Water Gate”. Carved on the gate were the Chinese characters of “Gongchen”. The present gate is modeled on the original.

南关清真寺

西宁著名的清真寺之一。位于城东区中南关 23 号，建于 1934 年，总占地面积为 3100 平方米。寺内有砖混结构礼拜大殿五间；二层砖混结构、高约 15 米的唤醒楼一座。该寺附近居住的信教群众 1600 余户。

Nanguan Mosque

Located at No. 23 Zhongnanguan in Chengdong District, it was built in 1934 and covers an area of 3100 square meters. Five worship halls with brick-concrete structure were built in the temple. A two-storey brick-concrete structure with a wake-up building about 15 meters high was built. There are more than 1600 religious families living near the temple. It is one of the famous mosques in Xining.

金塔寺

始建于明万历年间，系塔尔寺属寺，为塔尔寺驻西宁办事处。寺院为坐南朝北的四合院，占地面积666平方米，主供释迦牟尼佛。

Gold Roof Temple

Built during the reign of Emperor Wanli of the Ming Dynasty (1368-1644), it is affiliated with Kumbum Monastery in Xining. Its square courtyard of 666 km² is situated in the south facing north. The main Buddha image is that of Sakyamuni.

大佛寺

位于西宁市教场街东南端，始建于宋淳化元年（990 年），藏传佛教的古刹，是后弘期复兴的圣地。曾是西宁市城内的佛教四大寺院之一。其建寺时间之早、名声之大，远在西宁市的宏觉寺、经塔寺、专经寺之上。

Big Buddha Temple

Located at the southeast end of Jiaochang Street, Xining City, Qinghai Province, which was founded in 990, it is an ancient temple of Tibetan Buddhism and a holy place for the revival of later Hongqi Period. It was one of the four Buddhist monasteries in Xining.

位于西宁市文庙街，始建于明宣德三年（1428年），后经历了四次扩建修缮之后形成占地80余亩的规模。历经几百年的风雨，如今这规模宏大的古建筑群只剩下大成殿。2004年在原址修建了店铺、广场等设施。修缮后的文庙以现存的大成殿为中心，并和现在的文化街夜市相连接，成为西宁市城中区新的商业区。

Confucious' Temple

Located in the Confucian Temple Street, it was built in 1428. After four times of expansion and renovation, it has formed a scale of more than 80 mu. After hundreds of years of storms, now only the Grand Hall is left in this large-scale ancient building complex. In 2004, shops, squares and other facilities were built at the original site. The renovated Confucian Temple, centered on the existing Dacheng Hall and connected with the current night market of cultural street, has become a new commercial district in downtown Xining.

山陕会馆

位于西宁市城中区饮马街东侧的兴隆巷，始建于清光绪十四年（1888 年），最早建在今东关大街路北，后遭焚毁，光绪二十五年（1899 年），由山西、陕西商人再度筹资修建于现址。现存建筑规模宏大，布局严谨，殿宇华丽，结构精巧。是山西、陕西两省商贾联乡谊、祀神明的处所，集精巧的建筑结构和精湛的雕刻艺术于一身，充分显于了古代劳动人民的智慧与才能，是中国古代宫殿建筑的杰作。

Shaanxi Guild Hall

Xinglong Lane, located on the east side of Yinma Street in the middle of the city, was built in 1888. It was first built in the north of Dongguan Street and was burned down. In 1899, it was built by Shanxi and Shaanxi businessmen again. The existing buildings are huge, well-laid out, magnificent and exquisite in structure. It is the place where Shanxi and Shaanxi merchants and businessmen associate with each other and worship gods.

莫家街

位于西宁市中心东大街南侧，与饮马街相对，迄今已经有 600 年的历史了。莫家街夜市是市民、游客夜生活的好去处。

Mojia Street

Located on south side of Dongda Street in downtown Xining, opposite to Yinha Street, it is more than 600 years old. The nearby night market is very popular.

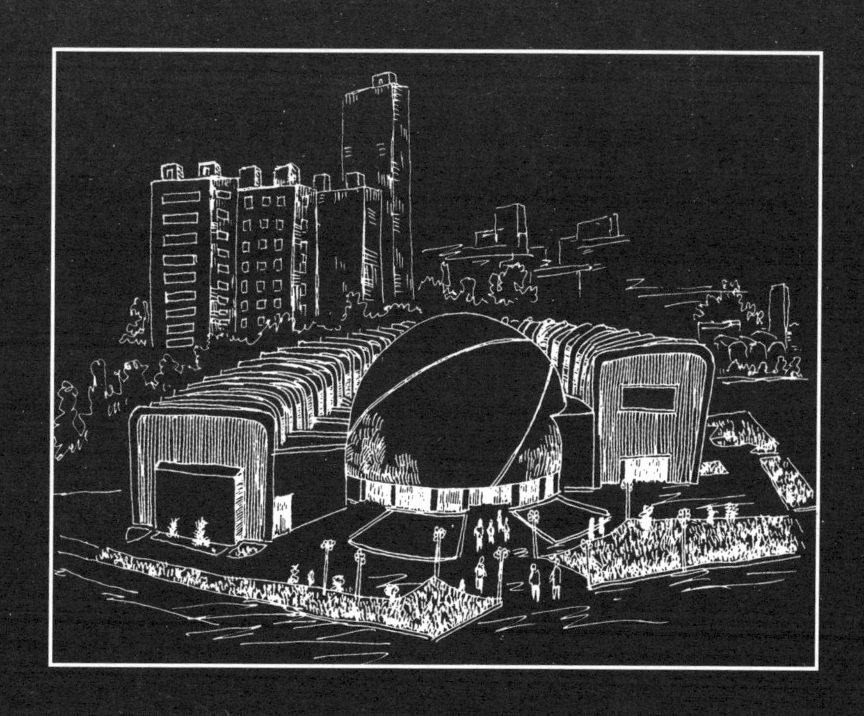

近现代建筑
Contemporary Architecture

湟源县城关第一小学

始建于 1920 年，距今有近百年历史，坐落在风景秀丽的丹噶尔古镇中心，至今保留着古朴厚重的建筑风格。

Chengguan 1st Elementary School, Huangyuan County

Built in 1920 in downtown Dange'r, this nearly 100-year old structure is simple yet delicate.

马步芳公馆·全景

位于西宁市城东区为民巷。马步芳，甘肃临夏人，为民国时期西北地区军阀马家军重要人物。公馆始建于 1942 年，为马步芳私邸，取名为“馨庐”。 在马公馆里许多建筑的墙面镶有玉石，故人们称其为“玉石公馆”。公馆由多个院落和不同形式的房舍以及花园组成，各个院落的房舍布置有序，结构严谨。构成了统一和谐的整体。

Panorama of Mabufang Mansion

Built on Weimin Lane, Chengdong District in downtown Xining, by a famous warlord in 1942 during the National Period (1912-1949), as his private residence, it is named" Xinlu"(Pleasant Home). There were several courtyards, houses, and gardens, each with its own specific design and structure.

马步芳公馆·玉石亭

马步芳天生痴迷玉石，在建造公馆时将大量昆仑玉石镶嵌在建筑物里。后该建筑被命名为玉石亭。

Yushi Pavilion

Ma Bufang loved jade so much that he used a lot of in the construction of the mansion; thus it is also called "Jade Mansion."

马步芳公馆·晓泉古井

位于公馆院内，此井与晓泉相同，是马步芳公馆主要饮用水源。该井水质甘甜，清澈见底，现已停用。

Xiaoquan Old Well, Mabufang Mansion

This old well located at the mansion and connected with Xiaoquan (Small Spring), used to provide sweet and clear drinking water for the entire mansion, but now it is closed.

马步芳公馆·壁炉

位于公馆贵宾厅右厢房中，为仿俄式方形壁炉，壁炉表面由玉石镶嵌，周围饰有花纹图案，做工精致，别具一格。

Fireplace of Mabufang Mansion

The fireplace is in the right side living room of the mansion. It is a Russian style square fireplace decorated with jade flower patterns. The design and structure are unique and delicate.

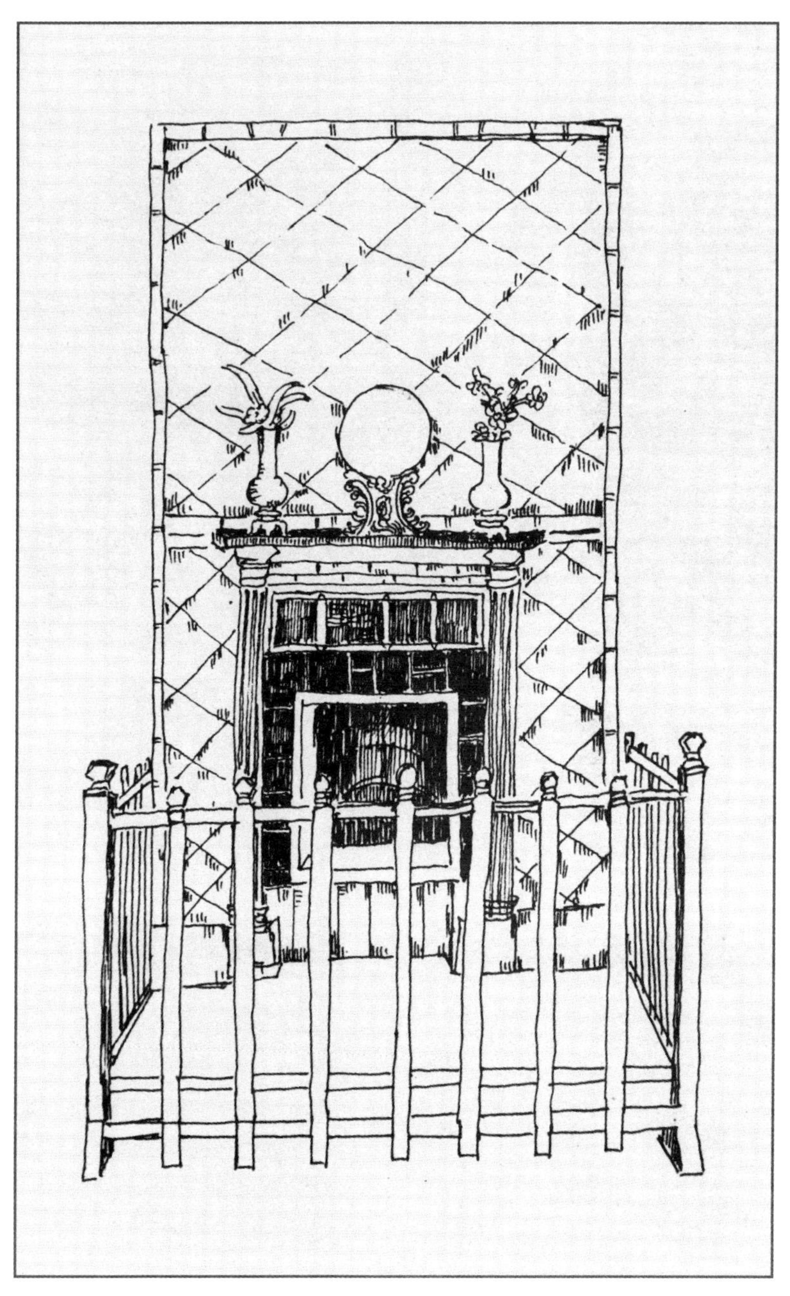

马步芳公馆·女眷楼

位于公馆建筑群的西南角，又叫南楼小院。可通花园，是古典回廊木结构的中式二层楼四合院，是女性宾客和部分女佣人住宿的地方，一楼是女佣住的，二楼是女宾住的，这个楼当年绝对禁止男人入内。

Ladies Building, Mabufang Mansion

Also called "South Courtyard" located on the southwest of the mansion complex, it is a Chinese style wooden residential building specifically built for female guests and housekeepers.

廖霭庭旧居

位于西宁市南玉井巷四号，建造于1945年。其主人廖霭庭是青海西宁人，对青海工商业有过突出贡献，曾任三原富民织布厂厂长，西宁市商会监察委员、常务委员。60年来，这座古民居始终在城市的喧嚣与繁华中，独守着一份宁静与从容。

Former Residence of Liao Aiting

Located at No.4 Nanyujin Lane in downtown Xining, it was built in 1945 by Liao Aiting who contributed much to the industrial and commercial development of Qinghai. For 60 years this house has remained peaceful and tranquil among the hustle and bustle of the city.

青海民族大学

创建于 1949 年，是青海最早建立的民族院校之一。有汉、藏、回、土、撒拉、蒙古等几十个民族的学生在校学习。

Qinghai University of Nationalities

Founded in 1949 as one of the earliest ethnic universities in Qinghai, it enrolls students from 39 ethnicities including Han, Zang, Hui, Tu, Sala, Mongolian.

青海师范大学

建于 1956 年，学校以本科教育为重点，是一所具有高原地域和民族特色的省属重点大学。

Qinghai Normal University

Built in 1956, the university mainly enrolls undergraduate students now. It is a provincial level university with local and national characteristics.

青海大学

前身为青海工学院，始建于 1958 年。1971 年成为包括工农两大学科在内的青海工农学院。1988 年更名为青海大学。

Qinghai University

Formerly Qinghai Technical University built in 1958, in 1988 its name was changed to Qinghai University.

青海省博物馆

是青海省地方历史、民族、民俗、宗教等文物的搜集和研究中心。如今馆藏各类文物已达 4.7 万余件，以实物的形式集中反映青海不同历史时期的发展概貌。

Qinghai Museum

It is the searching and researching center for Qinghai chronicles, including history, ethnicity, customs, religion, etc. Presently it houses 47 thousand exhibitions mostly reflecting the development of Qinghai over the centuries.

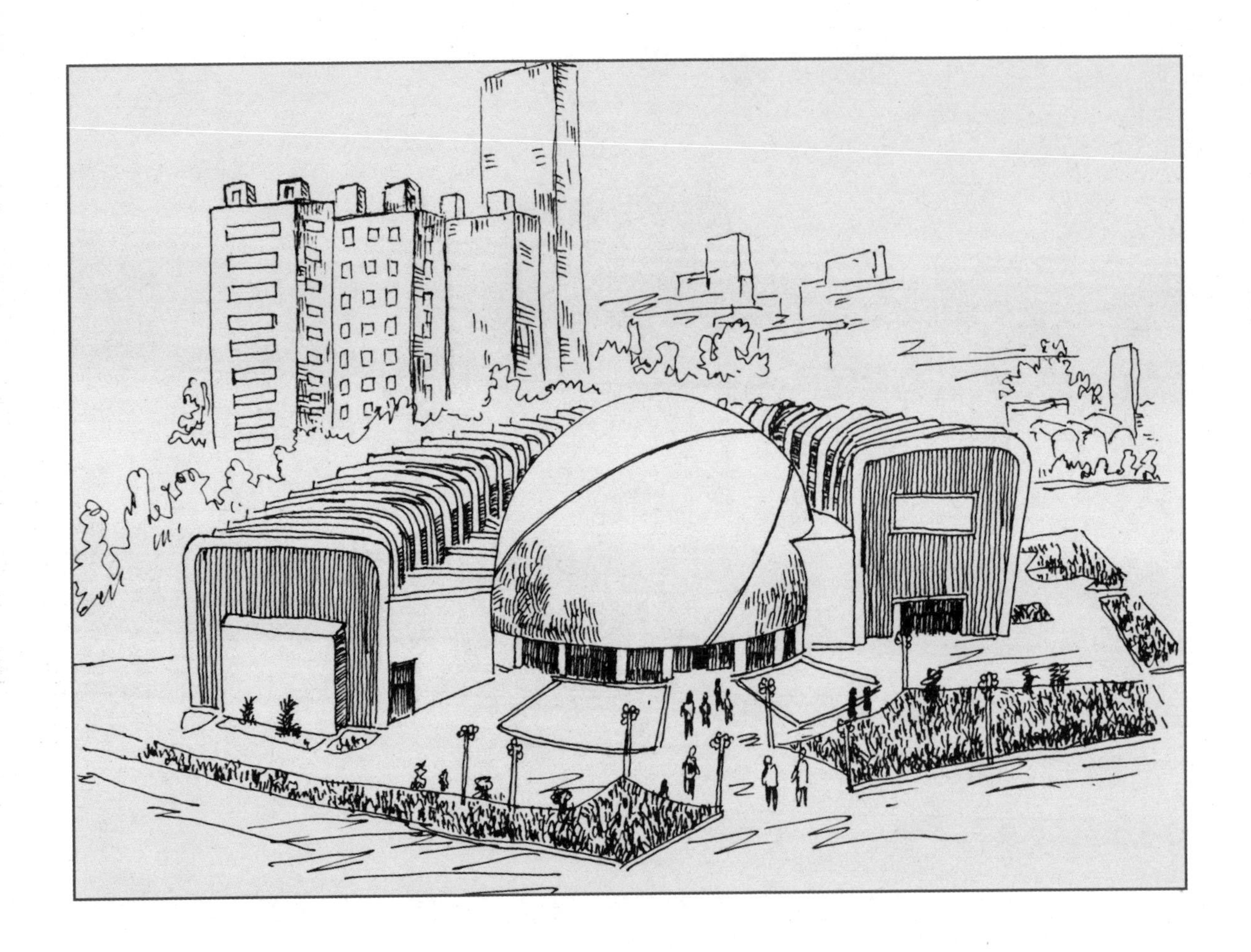

青海省科学技术馆

始建于 1984 年，以科技展览教育为主体，辅以科技实践活动和科技培训，是目前青海省最大的科普活动场所。

Science and Technology Museum of Qinghai

Built in 1984, the museum focuses on education in science and technology, supplemented with scientific and technological activities and training. It is the main venue for scientific and technological activity in Qinghai.

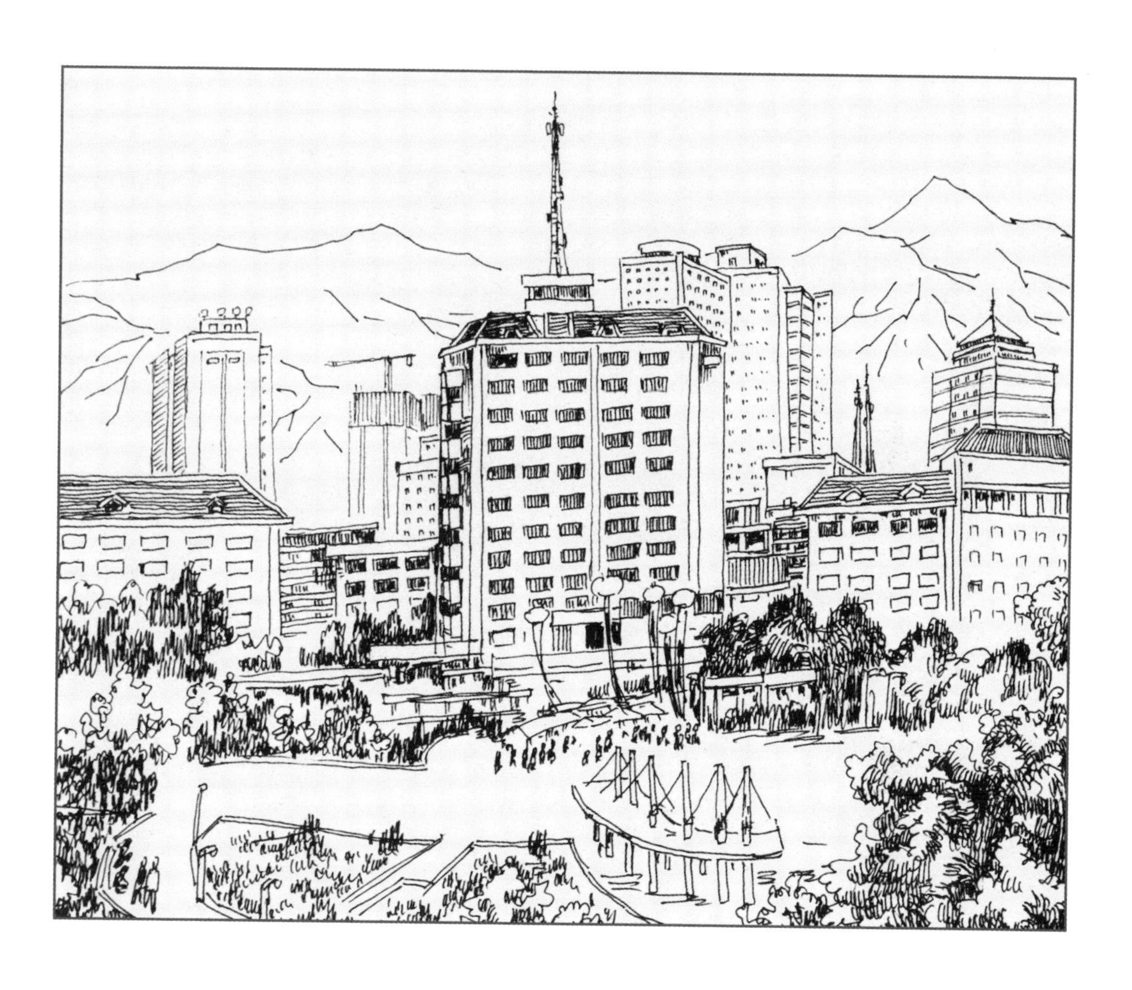

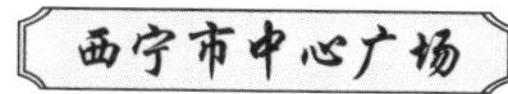

位于西宁市城市核心区，占地面积7万多平方米，是市民健身、娱乐，休闲的场所。

Xining Central Square

Located in downtown Xining, its 70,000 square meters are used for exercise, entertainment and relaxation.

新宁广场

位于西宁市城西区五四大街与新宁路交汇处，东接省博物馆，西联新宁路，北邻五四大街，南靠省图书馆。总面积 11.1 万平方米，绿地面积 6.6 万平方米。

Xinning Square

Located at the intersection of Wusi Street and Xinning Road in downtown Xining, it is adjacent to the Museum, and the Provincial Library.

青海体育运动中心

坐落于西宁市海湖新区，占地367亩，总建筑面积18.8万平方米。是我国西部地区规模最大、配套设施最完善的全民健身基地。

Qinghai Gymnasium

Located at Haihu New Zone in downtown Xining, with a total area of 188,000 m², it is the city's largest and most well-equipped gymnasium.

人民公园

位于西宁市城西湟水与北川河交汇处南岸，西宁最大的公共游乐公园，总面积 36 公顷。这里绿树成荫、曲径回廊、水阁相间、鸟语花香，十分清雅。

People's Park (1)

Located at the juncture Huangshui and Beichuan River in south Xining, it is the largest public park. It has zigzagged corridors and various pavilions built over the water.

文化公园

位于城西区海晏路，占地230亩。公园采用规则式和自然景观相结合的布局手法，现代特色突出，具有欧式风格。

People's Park (2)

Located at Haiyan Road, Chengxi District in Xining, this European style park is a combination of structures and natural landscaping.

海湖新区

西宁市海湖新区规划面积为 10.84 平方千米，新区突出高原城市特点，形成数个成规模的街头绿地及主题公园。

Haihu New Zone

Haihu New Zone will have an area of 10.84 km² manifesting the characteristics of the plateau, featuring several large green zones and theme parks.

青海藏文化馆

坐落于西宁市湟中县鲁沙尔镇莲花湖畔，是一个集趣味性、观赏性和参与性于一体的藏族文化场馆。

Tibetan Cultural Museum, Qinghai

Located beside Lotus Lake in Lushaer Town, Huangzhong County, Xining, the museum combines fun, appreciation and participation.

西宁体育馆

位于西宁市中心西门口西南侧的南川河畔，1982 年建成并投入使用。该馆占地面积四万平方米，这是一座高 35 米、底层长 68 米，宽 56 米，主体建筑面积 11064 平方米、可以容纳 8000 多名观众的体育建筑，是西北地区最大的体育馆之一。

Xining Stadium

Located on the south-west side of the west gate of Xining City, Qinghai Province, it was built and put into use in 1982. The gymnasium covers an area of 40,000 square meters. Its main building area is 11064 square meters. It can accommodate more than 8,000 spectators. It is one of the largest Gymnasiums in Northwest China.

老爷山

位于西宁市北约 40 千米，大通县境内，又名元朔山。登高眺望，云海苍茫，漫步林中，别有野趣。

Laoye Mountain

Also called Yuansu Mountain, located in Datong County, 40km north of Xining, it is a good place for mountain-climbing and hiking.

西宁街景

西宁是一个拥有悠久历史的高原古城，是中国黄河流域文化的组成部分。早在四五千年以前就有人类在这块土地上生产、生活，繁衍生息。北宋崇宁三年（1104 年）改为西宁州，至此“西宁”之称始于见史。1949 年 9 月 8 日成立市人民政府，为青海省辖市。

Streetscape of Xining

As an integral part of Yellow River culture, Xining is an ancient plateau city with a long history. During the Chongning reign of the North Song Dynasty(1104 AD), the city got its current name. On 8 September 1949 the People's Government of Qinghai was founded with Xining as its capital.

中国工农红军西路军烈士纪念馆

位于西宁市南川东路烈士陵园内，南川河畔，风景秀丽。陵园由烈士群雕塑像、纪念碑、烈士墓、中国工农红军西路军纪念馆等组成。

Martyrs' Memorial Hall of the Xilu Troop of Red Army

Located in Martyrs' Park, Nanchuang East Road, downtown Xining close to the Nanchuan Riverbank, the scenery is very beautiful. Landscapes include: Martyrs' Statures, Monuments, Cemeteries, Memorial Hall of Xilu Troop of Red Army, etc.

西宁烈士陵园

位于西宁市凤凰山下，建于 1954 年，园内安葬着 1776 位烈士遗骨。该馆是为缅怀在青海牺牲的中国工农红军西路军烈士而建立。

Martyrs' Cemetery of Xining

Built at the foot of Fenghuang Mountain (Phoenix Mountain) in 1954, it holds the remains of 1,776 martyrs. It is in remembrance of those martyrs of Xilu troop of Red Army who sacrificed their lives in Qinghai.

王洛宾雕像

坐落于西宁文化公园。王洛宾，民族音乐学家。1939 年定居西宁，创作了青海民歌《在那遥远的地方》，一生整理和创作了 700 余首西部民歌，被誉为“西部歌王”。

Wang Luobin Statue

The statue is located in Xining Culture Park. Wang Luobin was a folk musician who created the famous Qinghai ballad "In that remote place". During his lifetime he composed more than 700 ballads and was known as "*King of the Western Ballad*."

古老水车

水车又称孔明车，是我国最古老的农业灌溉工具，是先人们在征服世界的过程中创造出来的生产工具，是珍贵的历史文化遗产，至今已有 1700 余年历史。

Old Waterwheel

Also called Kongming Wheel, it is the most ancient irrigation device in agriculture created by our ancestors during the process of conquering the Nature. It is a precious historical and cultural heritage more than 1,700 years old.

藏族民居

以土木结构为主，内设天井、天窗等，较好地解决了气候、地理等自然环境不利因素对生产、生活的影响，达到了通风、采暖的效果。

Tibetan Residence

The houses mainly built of wood, with patio and skylight which can better solve the climatic and geographic problems, help keep the houses ventilated and warm.

民风民俗
Customs and Cultures

藏族宗教活动·转经

藏传佛教的一种活动，即围绕着某一特定路线行走、祈祷。

Tibetan Buddhism Activity · Prayer Wheels

Prayer Wheels are a popular religious activity among Tibetan Buddhists in Tibet, Sichuan, Yunnan, Qinghai and Gansu. Devotees walk and pray along specific routes.

藏族宗教活动·磕长头

“磕长头”是藏传佛教信仰者最至诚的礼佛方式之一，为等身长头，五体投地匍匐，双手前直伸。每伏身一次，以手划地为号，起身后前行到记号处再匍匐，如此周而复始。

Tibetan Buddhism Activity · Kowtowing

Kowtowing, an act showing respect and reverence by full prostration of the body, is a revered practice among Tibetan Buddhists. The worshippers move forward one body length at a time.

藏族宗教活动·辩经

辩论佛教教义的学习课程，是藏传佛教喇嘛攻读显宗经典的必经方式。多在寺院内空旷之地、树荫下进行，主要分为对辩和立宗辩两种形式。

Tibetan Buddhism Activity · Debating

Debating is a compulsory course for the lamas to study classics of Tibetan Buddhism. It is usually held in the open or under the trees. There are mainly two types of debate: one-to-one and sect-to-sect.

藏族宗教活动·摸顶赐福

藏传佛教高僧、活佛为僧俗信众赐福、消灾而举行的一种传统宗教仪式。在藏族传统文化里，只有活佛和高僧才能抚摸别人的头部并赐福。因此，得到活佛的摸顶对信众来说是一种莫大的荣幸。

Tibetan Buddhism Activity · Blessings by Touching Heads

Eminent monks and Living Buddhas bless believers by touching their heads. It is a traditional religious ritual of giving blessings and dispelling evil spirits.

藏族宗教活动·藏戏

藏戏起源于8世纪藏族的宗教艺术，17世纪时，从寺院宗教仪式中分离出来，逐渐形成以唱为主，唱、诵、舞、表、白和技等基本程式相结合的生活化的表演。藏戏唱腔高亢雄浑，基本上是因人定曲，每句唱腔都有人声帮和。藏戏原系广场剧，只有一鼓一钹伴奏，别无其他乐器。

Tibetan Buddhism Activity · Tibetan Opera

Tibetan Opera, Lhamo or Ache Lhamo in Tibetan Language, originated from the religious art. During the 17th century, it separated from monastic religion and gradually developed into a cultural art form of the common people. The voice of Tibetan opera is high pitched and fulsome, accompanied only by drums and cymbals.

藏族宗教活动·军舞

大型广场歌舞，类似于中原汉族聚集地的社火和西藏的“望果节”表演，具有广泛的群众性和娱乐性，舞蹈风格雄壮、欢快，从中表现了藏族人民向往美好、安康生活的愿望。

Tibetan Buddhism Activity · Army Dance

Large scale square dances of the Zang people combine fellowship and entertainment. The square dance is energetic and joyful, demonstrating a wish for better and healthy lives.

藏族宗教活动·奏乐

藏族传统音乐特色鲜明，品种多样，包括民间音乐、宗教音乐、宫廷音乐三大类。三大方言区的民间音乐在风格上有明显的差别，乐种亦不尽相同。

Tibetan Buddhism Activity · Music

The traditional Tibetan music is distinct and diverse. It includes folk, religious and court music. The music of the three main dialect zones are different in style and music type.

藏历新年

藏族人民共同的传统节日，寺僧和俗人一样也欢庆一年一度的新春佳节。不过藏历年的推算法与农历春节有些差异。

Tibetan New Year

The traditional Tibetan new year is based on the Tibetan calendar, not the Chinese.

雪顿节

每年藏历六月底七月初举办，是西藏传统的节日。传统的雪顿节以展佛为序幕，以演藏戏看藏戏、群众游园为主要内容，同时还有精彩的赛牦牛和马术表演等。

Tibet Shoton Festival

This is a tradition gala in Tibet held in late June and early July of Tibetan calendar. The main activities consist of Buddha image displays, Tibetan opera performance and group entertainment activities. There are also activities like yak and horse racing.

插箭节

藏族民间流传的古老的由祭祀仪式衍化而成的节日，在藏区普遍保留的插箭节仪式，实际上就是祭祀、祈求“保护神”保护的仪式。

Interpolation of Arrow Festival

Derived from traditional religious rituals of Tibetan Buddhism, it is still common in Tibet, intending to give offerings or seek protection from the gods.

赛马节

在所有的民间传承的藏族节日中，几乎都少不了赛马活动，并且此项活动有着悠久的历史。在藏族的节日民俗中，赛马常以主题的形式在节日中显现，而且更为重要的是，建立在对马的浓郁感情基础之上的藏族人民，创造了独具民族特色的赛马文化。

Horseracing Festival

Horseracing, a traditional Tibetan folk activity, is still very popular during almost every Tibetan festival. People have developed special feelings towards horses throughout the years and created a unique national horseracing culture.

塔尔寺四大观经会

塔尔寺每年举行四次大型法会和两次小型法会。在四大法会期间，最隆重盛大的佛事活动是“跳神”“晒大佛”和大型酥油花展。

The Four Religious Events of Kumbum Monastery

Every year at Kumbum Monastery there are four large and two small religious events. In them the most important Buddhist activities are "Lamaist Devil Dance", "Displaying the Buddha" and large displays of butter sculptures. Local people call them "The Four Major Events."

晒佛节

藏族人民的传统宗教节日，大都在藏历二月初、四月中旬或六月中旬举行，具体日期各地不尽相同。届时，各地寺庙将寺内珍藏的巨幅布画和锦缎织绣佛像取出，或展示于寺庙附近晒佛台，或山坡，或巨岩的石壁之上供群众观瞻。

The Buddha Display Festival

This is a traditional Tibetan religious festival in early February, mid April or Mid June of Tibetan calendar. During the festival, the monasteries will move the images of the Buddha from the temples to "displaying terraces" near the temples, on the slopes of the mountains or on the stone walls for people to view and honor.

六月欢乐节

是同仁县藏族、土族村庄特有的传统文化节，已流传1400多年。这是一种原始宗教氛围浓烈、文化形态与文化内涵复杂而丰富的宗教性节日。热贡地区的藏族、土族都会参加。

Joyous June Festival

This is a traditional religious and cultural festival of the Zang and Tu ethnic groups in Tongren County. It has been celebrated for more than 1,400 years.

锅庄舞

又称为“果卓”“歌庄”等，藏语意为圆圈歌舞，是藏族三大民间舞蹈之一。

Guozhang Dance

Also called “Guozhuo” or “Gezhuang”, in Tibetan language it means “ Circle Dance”. It is one of the three major Tibetan folk dances.

经 幡

又称风马旗，或音译为隆达、龙达，是指在藏传佛教地区的祈祷石或寺院顶上、敖包顶上竖立的以各色布条写上的八字真言、六字真言等经咒，捆扎成串，用木棍竖立起来的旗子。因布条上画有风马一只，寓意把祷文借风马传播各处，故得名“风马旗”。

Streamers

Also called "Racing Horse Flags", in Tibetan Buddhism areas people write mantras on these colored cloths, bundle them together into a cluster and put them up with wooden sticks. A racing horse is usually drawn on them symbolizing spreading of the mantra can be as fast as a racing horse, thus the name " Racing Horse Flag."

青海社火·历史

青海汉族大都是历代从内地迁徙而来的，也把内地传统的社火带到了青海，同时又融合了当地民族文化的特点，于是出现了具有地方特色的社火活动。从正月初七开始闹社火，丰富多彩的表演形式在青海各民族中广为流传。

Qinghai Festive Activities · History

The Han people in Qinghai are mostly decedents of those who immigrated from eastern China during different dynasties. When they moved to Qinghai, they brought their traditional of festive activities but also integrated them with the local culture, thus forming today's Festival Activities with local characteristics.

青海社火·旱船

旱船是中国民间表演艺术形式之一，在陕西、山西、甘肃境内各地都广为流行，这是一种模拟水中行船的汉族民间舞蹈。青海社火里的旱船表演，在保留原有旱船表演的形式外更糅合了青海各民族的一些民俗文化和艺术形式，形成了丰富多彩且独具高原特色的社火艺术。

Qinghai Festive Activities · Land Boat Dancing

Land Boat Dancing is a folk dance mimicking boating on the water. Land boat dancing in the Qinghai area has kept the original form but integrated folk culture and art forms of different Qinghai ethnic, groups thus forming a colorful festival activity with plateau characteristics.

青海社火·高跷

高跷也叫“高跷秧歌”，是一种广泛流传于全国各地的民间舞蹈，因舞蹈时多双脚踩踏木跷而得名。高跷是社火的表演形式之一，深受青海人民的喜爱。

Qinghai Festive Activities · Stilt Dance

Stilt dance also called “ Stilt Yange”, is a nationwide folk dance in China. It got its name because participants dance on stilts. It is one of the popular performances during the festivals loved by the people in Qinghai.

青海土族波波会·跳神

“波波会”是土族传统的民俗活动。“波波”为土族语，意为法师作道场，俗称“跳神”，是一种驱邪仪式。跳神是用黄表和彩纸剪贴的云纹、万字纹等花样长幡，挂在杆头，垂落于地，幡杆顶端横置两齿叉，叉尖各戳一个大馒头在村庄跳舞驱邪。

TuBobo · The Devil Dance

Event Bobo Event is a traditional Tu Nationality folklore activity. Bobo is the language in Tu, meaning “sorcery “or “expel the evil spirit”. Usually a long paper-cut streamer is hung around a wood stick. On the top of the stick a big bun is thrusted there. With the stick in his hand, the sorcerer then dance around in the village to expel the evil spirits.

青海土族波波会·招魂

“波波会”的高潮是做道场，人们把所有供品拿到广场上，上香、磕头祷祝。然后由大法师领班，其余法师随其后，手举法鼓，齐敲鼓点，高颂祷词，左族右转，前移后挪舞动祈福。

TuBobo · Calling Back the Spirit of the Dead

The peak of Bobo Event is chanting Dojo. People bring all their offerings to the square, light incense, bow their heads and pray. Then the team start to dance and pray for blessings led by the Exorcist, followed by other Taoism priests.

青海土族波波会·轮子秋

土族传统娱乐项目。每年正月农忙过后，土族青年们将大板车改装成融秋千、转盘为一体的吊车，两名身着彩装的姑娘在上面飞旋起舞。为求一年神清气爽，男女老幼在“轮子秋”上转一转已成习俗，技艺高超者还能做出高难度的杂技动作。

TuBobo · Wheels of Autumn

This is a traditional Tu entertainment activity. Every January after all farm work is done, young Tu people usually transform their handcarts into crane-like vehicles resembling the swing and turn plates. Then two young girls in colorful costumes dance on it.

青海土族波波会·法事

每到“波波会”时节都要法师做道场，做法驱邪祈福。

TuBobo · Dojo

Every year in Bobo Event sorcerers chant dojo to expel evil spirits and pray for blessings.

青海花儿·对唱

“花儿”是流传在青海广大地区的民歌。青海是花儿的故乡，聚居在青海东部地区的回、汉、撒拉、土等民族，各自创造了独具特色的“花儿”歌曲。它内容丰富多彩，形式自由活泼，语言生动形象，曲调高昂优美，具有浓郁的生活气息和乡土特色。

Qinghai Flowery · Pair Chanting

" Flowery" refers to a kind of folk song popular in Qinghai. Qinghai is the home of Flowery. Hui, Han, Sala and Tu folk have all created their own unique types of Flowery songs.

青海花儿·舞台表演

“花儿会”的习俗，一般在农历四、五、六月间，群众聚集在一起放嗓对歌，表演方式多样。

Qinghai Flowery · Dancing on Stage

The Flowery Event is a folk activity usually held in April, May and June during which people gather to sing and perform.

回族节日·开斋节

回历九月一日至十月一日为斋月，十月一日为开斋节，开斋节也叫尔德节。流行于在信仰伊斯兰教的民族中。

Hui Nationality Festival · Eid al-Fitr

Muslim Ramadan runs from 1 September to 1 October, the day ending Ramadan. It is a festival celebrated among the ten Islamic ethic groups in China.

回族节日·欢乐舞蹈

回族舞蹈大多是结合花儿表演进行，特点是模拟性强，动作干脆、细腻。回族传统的民间舞蹈有《巧看牡丹》《试全脚》《花儿式舞姿》等。

Hui Nationality Festival · Joyful Dance

Hui folk dance is usually performed together with Flowery, characterized by high-level mimicry and straight-forwardness.

小吃一条街

西宁是多民族聚居的城市，以回、藏、撒拉族为主，当地餐饮在西北地区颇有名气。西宁特色小吃给许多去西宁的游客留下难忘的回忆。

Snack Street

Xining is a multi-ethnic city, with Hui, Zang and Sala being the most numerous. The local restaurants are very famous in northwest China.

烤羊肉串

西宁牛、羊肉的肉质好，烤羊肉串是各地游客与当地群众喜爱的小吃美食。

Shish Kebab

Beef and goat meat are of excellent quality in Xining. Shish Kebab is a very popular snack for tourists and local folks.

青海湖自行车国际环湖赛

环湖赛从 2002 年开始，每年七八月在青海省举行。经国际自行车联盟批准，环湖赛为 2.HC 级，是亚洲顶级赛事，也是世界上最高海拔的国际性公路自行车赛。沿途自然风光雄奇壮美，旖旎迷人。

Tour of Qinghai Lake

The Bicycle Racing Tour of Qinghai Lake began in 2002. It is now held in July and August every year. The race is sanctioned by the International Bicycle Union (UCI) as a 2.HC race as part of UCI Asia Tour. It is one of the important bicycle road races in Asia and is held at the highest altitude in the world.

青海自然风光
Natural Scenery

祁连山

中国西北部重要山脉，是黄河和内陆水系的分水岭。祁连山的四季从来不甚分明，春不像春，夏不像夏。所谓“祁连六月雪”，就是祁连山气候和自然景观的写照，景色数不胜数。

Qilian Mountain

Qilian Mountain is an important mountain range in northwest China and the dividing range between the Yellow River and the inland water systems. The climate, diverse and complex, has formed beautiful natural landscapes.

青海湖

藏语名为“措温布”，意为“青色的海”。位于青藏高原东北部、青海省境内。是中国最大的内陆湖、咸水湖。由祁连山脉的大通山、日月山与青海南山之间的断层陷落形成。每年的 7-8 月是青海湖最美的时候。

Qinghai Lake

Qinghai Lake (Blue Sea Lake) is in Qinghai Province, northeast of the Qinghai-Tibet Plateau. It is the largest interior and the largest salt water lake in China. In Tibetan it means “ Blue Sea”. The scenery is most beautiful in July and August.

鸟 岛

在青海湖中，因岛上栖息数以十万计的候鸟而得名。鸟岛由两座小岛组成，西边小岛叫海西山，又叫小西山，也叫蛋岛；东边的大岛叫海西皮。鸟岛之所以成为鸟类繁衍生息的理想家园，主要是因为它有着独特的地理条件和自然环境，这里地势平坦，气候温和，三面绕水，是鸟类繁衍生息的天然场所。

Bird Island

So named because of the thousands of migrating birds visiting here, the island is an ideal place for birds to nest because of its unique rocky cliff geographic conditions and natural environment. The climate here is warm and the site is surrounded by water in three directions.

日月山

位于西宁市湟源县西南。风光秀美，而且具有重要的地理意义和历史意义。它是青海农区和牧区的分界线，海拔 3520 米，是游人进入青藏高原的必经之地，故有“西海屏风”“草原门户”之称。据说，文成公主当年入藏途经此山，因此也成为唐朝和吐蕃进行物资交流和两地使者往来的中转站。

Riyue Mountain

Riyue Mountain (Sun and Moon Mountain) is located southwest of Huanyuan County, Xining. The scenery here is beautiful and has very vital geographic and historical significance. It is the dividing line for agriculture and animal husbandry in Qinghai and it was also the intermediate trading and message station between the Tang China and Tibet.

北　山

青海互助北山国家森林地质公园坐落于祁连山脉南麓的东端，平均海拔 1800 米左右，可谓冬无严寒、夏无酷暑。山的南麓遍布茂密的原始森林，绵延数百公里。

North Mountain

Qinghai Huzhu North Mountain National Forest and Geographic Park is located at the eastern end of the southern slope of Qilian Mountain, with an average altitude of about 1,800m. On the south slope of the mountain are exuberant primitive forests extending hundreds of kilometers.

孟达天池

位于青海省东部循化撒拉族自治县东部，被誉为“青藏高原上的西双版纳”。孟达天池面积约 300 亩，蓄水量为 300 万立方米，平均水深为 17 米，海拔为 2504 米，周围分布有野生植物约 40 种，野生动物 60 余种。

Mengda Heaven Pool

Located at east Xunhua Sala Autonomous County, Qinghai Province, it is known as Xishuangbanna in the Qinghai-Tibetan Plateau. Mengda Heaven Pool has an area of about 300 acres, with an average altitude of 2,504m. There are more than 537 species of wild plants and more than 60 wild animal species in the area.

贵德丹霞

贵德丹霞地貌，形成于1.2亿年前。由水平或平缓的层状铁钙质混合不均匀胶结而成的红色碎屑岩构成，受垂直或高角度节理切割，在风化、重力崩塌、流水溶蚀、风力侵蚀等综合作用下形成形状各异、高耸入云的山崖，红似火，青如黛。

Guide Danxia

Danxia landform in Guide was formed 120,000,000 years ago. After millennia of weathering and erosion, it looks as red as fire and as blue as indigo.

贵德地质公园

位于贵德县境内，是以自然地貌景观和地质遗迹为主要特征，辅以多样生态景观和丰富人文景观的一个综合性地质公园。多种多样的地质遗迹反映了地质历史时期青藏高原的演化过程，也记录了黄河的发育史和贵德自然环境的变迁。

Guide Geographic Park

Located at Guide County. It is a comprehensive park characterized by natural landscapes and relics. There are also some diverse natural and man-made landscapes.

高山草原

重要的畜牧业基地，海拔 4000 米以上，称为高山草原。草地中星罗棋布地点缀着无数小湖泊，湖水清澈见底，游鱼可见。此外，草原植被还蕴藏着许多药用植物，可采收利用。

The Meadow on the Plateau

The alpine meadow is an important animal husbandry base. Since it is located 4,000m above sea level, it is called "Meadow on the Plateau". Dotted on the meadow are numerous small clear-water lakes.